Copyright © 2021

All rights reserved. No part of this publication may be reproduced, distributed, or transmitted in any form or by any means, including photocopying, recording, or other electronic or mechanical methods, without the prior written permission of the publisher, except in the case of brief quotations embodied in critical reviews and certain other noncommercial uses permitted by copyright law.

Thank you for your recent purchase. We hope you love it! If you do, would you consider posting an online review? This helps us to continue providing great products and helps potential buyers to make confident decisions. Thank you in advance for your review and for being a preferred customer.

This Book Belongs To

All Illustrations Are Hand Drawn

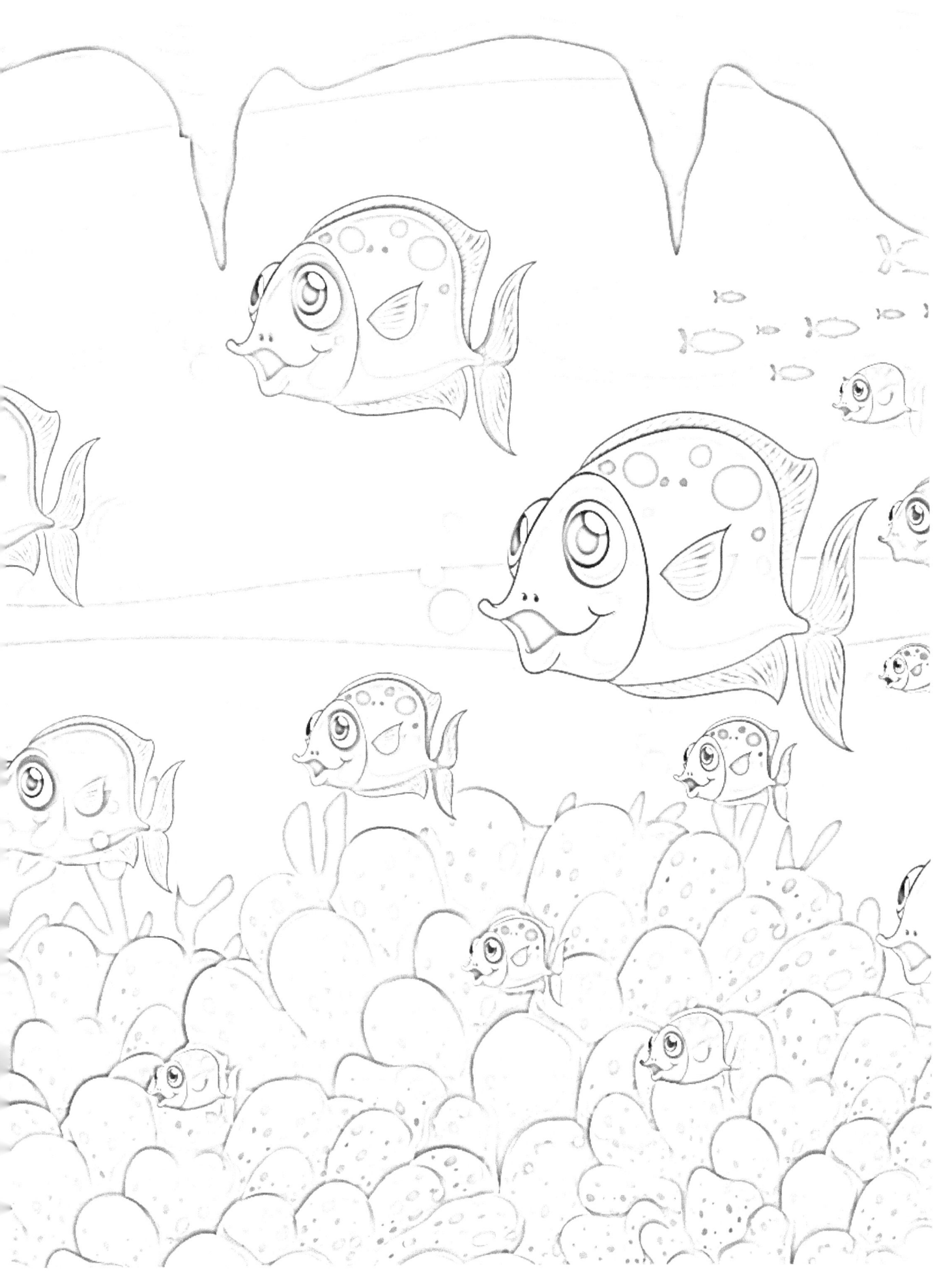

www.ingramcontent.com/pod-product-compliance
Lightning Source LLC
Chambersburg PA
CBHW080518220526
45465CB00006B/2524